MISSION CONTROL

THE JAMES WEBB SPACE TELESCOPE

JENNIFER LOMBARDO

PowerKiDS press

Published in 2025 by The Rosen Publishing Group, Inc.
2544 Clinton Street, Buffalo, NY 14224

First Edition

Editor: Jennifer Lombardo
Book Design: Rachel Rising

Photo Credits: Cover, p. 6 Dima Zel/Shutterstock.com; cover, pp. 1, 3–32 KK.KICKIN/Shutterstock.com; p. 4 Alexandre. ROSA/Shutterstock.com; p. 5 Felix Lipov/Shutterstock.com; p. 7 https://en.m.wikipedia.org/wiki/File:HST-SM4.jpeg; p. 8 24K-Production/Shutterstock.com; p. 9 https://en.m.wikipedia.org/wiki/File:James_Webb_Space_Telescope_2009_top.jpg; p. 10 LeonART Studio/Shutterstock.com; p. 11 Smart Black Box/Shutterstock.com; p. 12 FOTOKITA/Shutterstock.com; p. 13 NASA/ESA/M. Livio & Hubble 20th Anniversary Team (STScI); p. 14 dabyki.nadya/Shutterstock.com; p. 15 NASA/JPL-Caltech; p. 16 Vadim Sadovski/Shutterstock.com; p. 17 BEST-BACKGROUNDS/Shutterstock.com; p. 18 PeopleImages.com - Yuri A/Shutterstock.com; p. 19 Jbruiz/Shutterstock.com; p. 19 (background) A-Star/Shutterstock.com; p. 20 Moshe EINHORN/Shutterstock.com; p. 21 NASA, STScI; p. 22 https://en.m.wikipedia.org/wiki/File:NASA%E2%80%99s_Webb_Reveals_Cosmic_Cliffs,_Glittering_Landscape_of_Star_Birth.png; p. 23 https://en.m.wikipedia.org/wiki/File:Webb%27s_First_Deep_Field_(adjusted).jpg; p. 25 https://zh.wikipedia.org/wiki/File:Southern_Ring_Nebula_(NIRCam_and_MIRI_Images_Side_by_Side).png; p. 26 muratart/Shutterstock.com; p. 27 NASA, ESA, and the Hubble Heritage Team (STScI/AURA); p. 28 Triff/Shutterstock.com; p. 29 NASA.

Library of Congress Cataloging-in-Publication Data
Names: Lombardo, Jennifer, author.
Title: The James Webb Space Telescope / Jennifer Lombardo.
Description: Buffalo, NY : PowerKids Press, [2025] | Series: Mission control | Includes bibliographical references and index.
Identifiers: LCCN 2024036050 | ISBN 9781499449822 (library binding) | ISBN 9781499449815 (paperback) | ISBN 9781499449839 (ebook)
Subjects: LCSH: James Webb Space Telescope (Spacecraft)--Juvenile literature. | Space telescopes--Juvenile literature.
Classification: LCC QB500.269 .L66 2025 | DDC 522/.2919--dc23/eng/20240904
LC record available at https://lccn.loc.gov/2024036050

Manufactured in the United States of America

CPSIA Compliance Information: Batch #CWPK25. For further information contact Rosen Publishing at 1-800-237-9932.

CONTENTS

TELESCOPES AND OBSERVATORIES

People have been trying to get a better look at space for thousands of years. The first people to do this using glass lenses were Arab scientists in the 10th century. These scientists also made buildings called observatories where they could do their research. Around the 12th century, the Arabs' research made its way to Europe. A man named Hans Lippershey used it to make one of the first telescopes in 1608. A year later, Galileo Galilei improved Lippershey's design. Over time, other scientists continued to improve telescopes. Observatories were built all over the world, and as **technology** advanced, telescopes continued to get more powerful. However, they were all limited by one thing: Earth's atmosphere.

NASA KNOWLEDGE

Lippershey's invention was known as the perspicillum at first. The word "telescope" was invented in 1611. A Greek man named Giovanni Demisiani combined two parts of Greek words: *tele*, which means "far," and *skopos*, which means "see."

One of the most important observatories was built in present-day Uzbekistan by an Arab ruler named Ulugh Beg. Astronomers used what was visible through this hole to correctly measure things such as the length of a year, Earth's angle, and the locations of stars at certain times of the year.

A NEW KIND OF TELESCOPE

When we look up at the night sky, the stars often seem to sparkle. This is because we are seeing them through the layers of gas that make up Earth's atmosphere. In addition to **distorting** what we can see, the atmosphere blocks some **wavelengths** of light. This hides many objects from view, even when people use powerful telescopes.

By the 1940s, astronomers had realized that they would only be able to learn more about space if they could see past the atmosphere. They started studying and planning so they could build a telescope to launch into space. In April 1990, the project was completed. The telescope was named after astronomer Edwin Hubble. Since that time, the Hubble Space Telescope has been taking pictures and measurements of stars, planets, and other heavenly bodies.

AVOIDING LIGHT POLLUTION

Another advantage of putting a telescope in space is the total darkness. In cities, people can see very few stars—sometimes none at all. This is because the light from buildings, cars, and streetlights blocks the light the stars give off. More stars are visible in darker places, but because they are far away from most humans, those places are generally difficult to get to. They include the Sahara Desert, Siberia, and the Canary Islands.

The Hubble Space Telescope is 43.5 feet (13.2 m) long, and its widest point is 14 feet (4.3 m).

THE NEXT GENERATION

After seeing results from Hubble, astronomers realized that they wanted to learn more than the telescope was able to tell them. They started work on a new telescope, named the James Webb Space Telescope (JWST), in 2004. After completing it in 2016, NASA ran many tests on the new telescope to make sure everything worked properly. Finally, it was ready to launch in 2021.

Because Hubble was still working well, NASA scientists did not want to make the JWST **identical**. The point of the new telescope was to fill gaps in scientists' current knowledge, not to replace one telescope with the other. For this reason, the JWST is similar to Hubble, but it has different instruments, or tools, that help it measure different things.

NASA KNOWLEDGE

James E. Webb led NASA in the 1960s. He oversaw the moon landing and helped make NASA a world leader in astronomical research. Webb worked hard to get politicians to set aside money for NASA's research, successfully convincing them of the importance of space exploration.

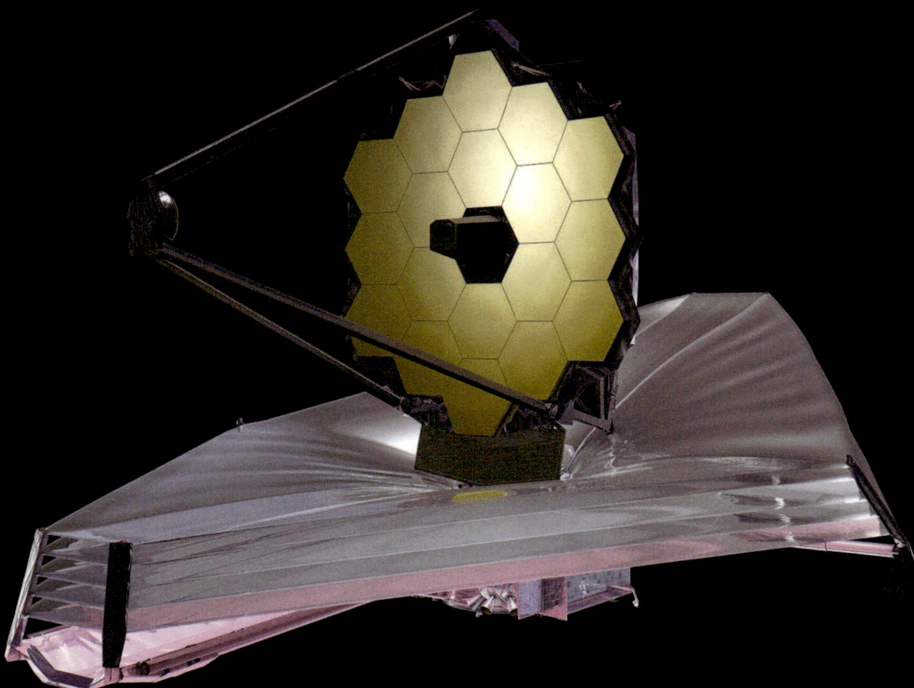

The JWST cost $10 billion to build. As of 2024, it is the world's largest and most powerful space telescope.

HUBBLE VS. WEBB

Light comes in many different wavelengths. Humans can only see a small part of the light **spectrum**, also called the electromagnetic spectrum, with the naked eye. We call the part we can see "visible light." Hubble was made to see mainly visible and ultraviolet light. However, the farther away an object is, the more infrared light it gives off. The JWST was made to see mainly infrared light. This allows it to see much farther than Hubble.

The JWST's main mirror is also much larger than Hubble's. This allows it to collect more light. By doing so, the JWST can see farther into the past than Hubble. This will help astronomers learn more about how the universe was formed.

LIGHT SPECTRUM

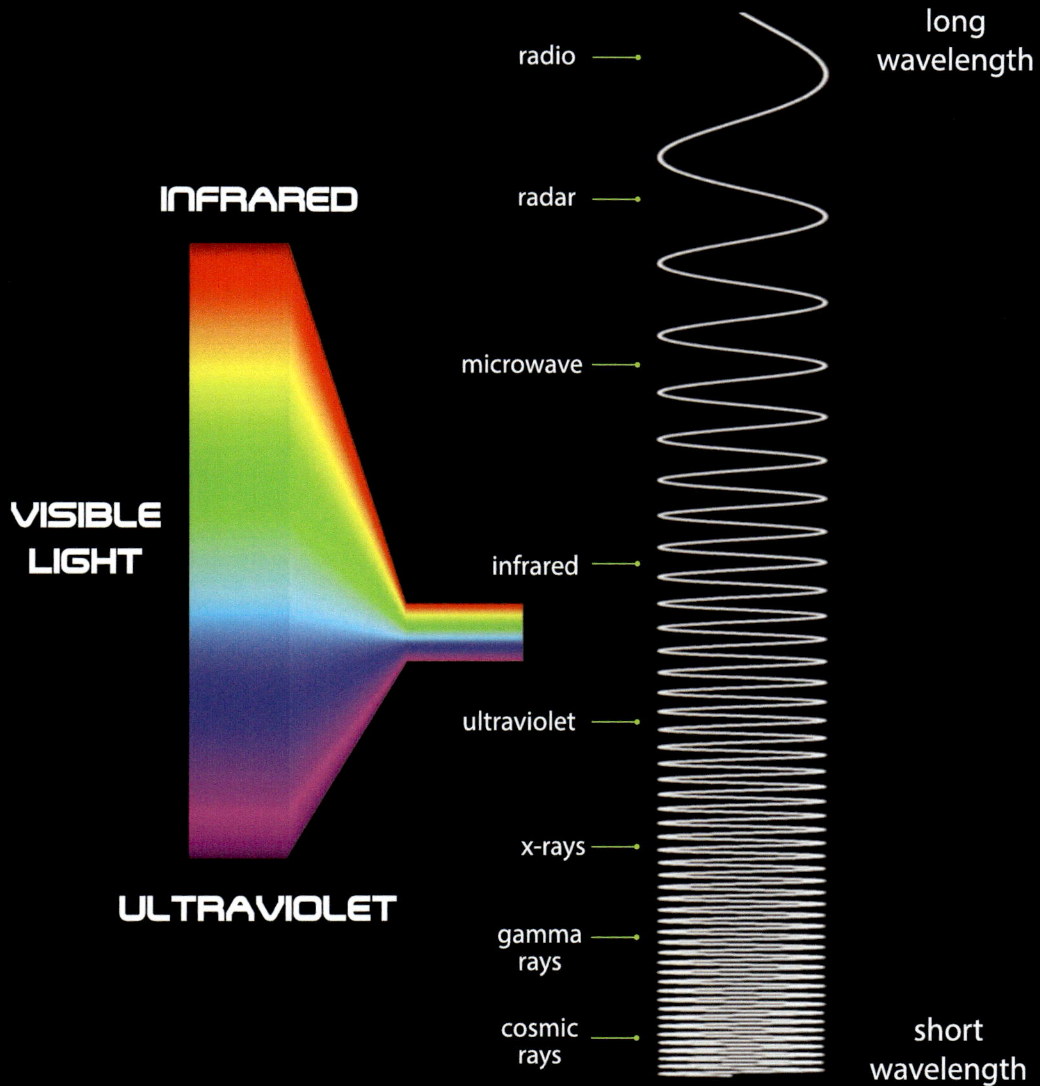

This picture shows the size and frequency of different waves in the electromagnetic spectrum.

SEEING THE PAST

How can a telescope look into the past? The answer has to do with the speed of light. Light travels about 186,000 miles per second (300,000 km/s). The farther away something is, the longer it takes the light to reach our eyes. When it does, what we see is how the light looked when it first started moving.

For example, imagine an alien planet 65 million light-years away from Earth. If the creatures living there could see the light from our planet, they would see Earth as it looked 65 million years ago. If they had telescopes powerful enough to let them see the planet's surface, they would be looking at dinosaurs! However, not even the JWST is that powerful.

NASA KNOWLEDGE

A light-year is a measure of distance, not time. It is the distance light travels in one year. Since light travels so quickly, it would take many times longer for a person to travel that same distance. Although spaceships in movies sometimes travel at light speed, no real spacecraft can do that.

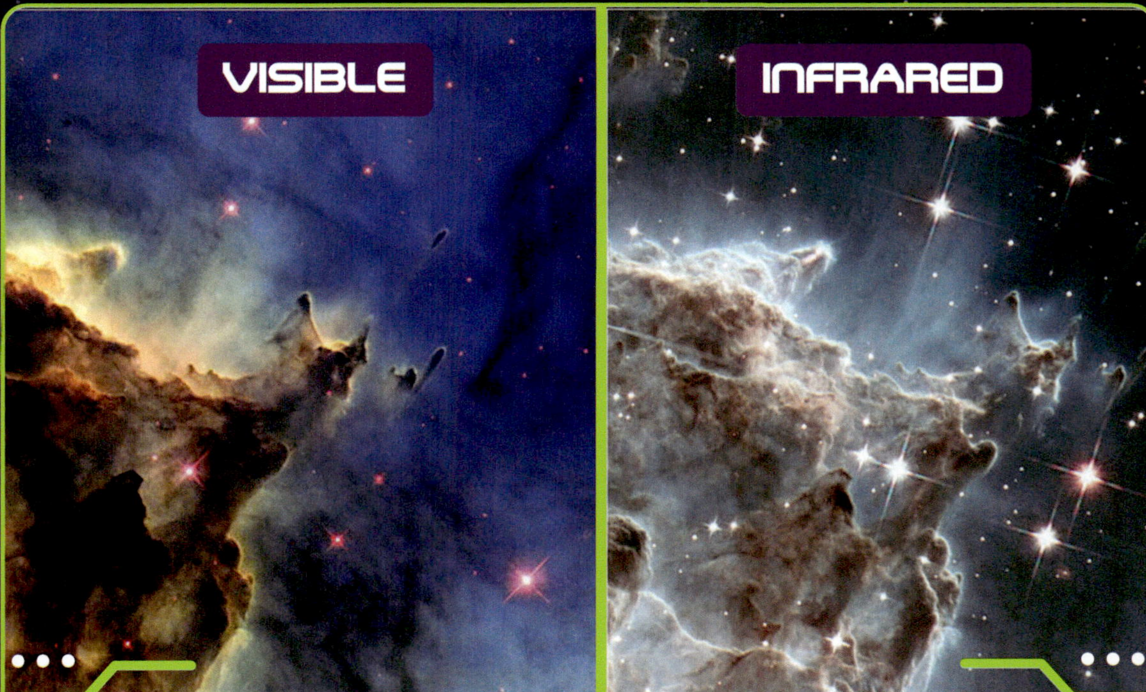

Shown here is the Monkey Head **Nebula** in both visible light (left) and infrared light (right). The infrared photo clearly shows stars that are not visible to the naked eye. This is because the visible light from newly formed stars is blocked by clouds of dust, but the infrared light is not.

HOW DOES IT WORK?

Both Hubble and the JWST are reflecting telescopes. This means they use mirrors to focus the light they take in. The kind of reflecting telescope people use in backyards and observatories has two mirrors in it. A large, curved mirror is placed at the bottom of the inside of the telescope. Light enters through the end opposite the mirror. The mirror bounces back, or reflects, the light.

Because a curved mirror can make images appear to be upside down, a reflecting telescope has a second, smaller mirror inside it. It is placed at an angle, so it reflects the image right-side up into the eyepiece instead of back out into space.

A BETTER TELESCOPE

The other kind of telescope that is commonly used on Earth is called a refracting telescope. This kind uses glass lenses instead of mirrors. Although both kinds work well, mirrors have more advantages than lenses for larger telescopes. The main reason for this is that the bigger a lens gets, the thicker it needs to be to work properly. This also makes the lens heavier. In contrast, a big mirror is just as powerful whether it's thick or thin.

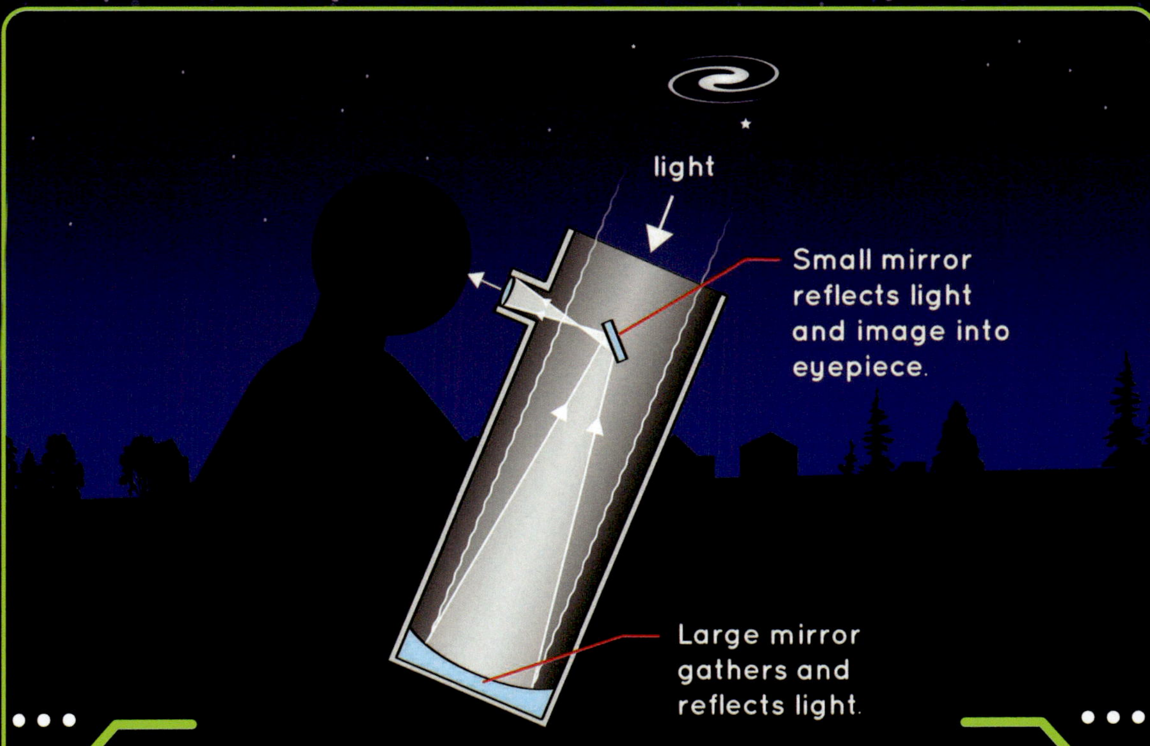

This picture shows how a reflecting telescope directs the light that passes through it.

BUILDING JWST

The JWST's first and largest mirror is made up of 18 hexagons fitted together. Its builders did this so the mirror could fold up, making it easier to launch. They used hexagons so the pieces would fit together well, with no gaps between them. Each hexagon is made of a lightweight metal called beryllium and covered with a thin layer of gold. The gold helps the mirror reflect infrared light more easily.

The primary mirror reflects light onto the secondary mirror. This is a much smaller mirror that directly faces the primary mirror. A foldable **tripod** holds the secondary mirror far enough away for the light to be reflected back into an opening in the center of the primary mirror. Each leg of the tripod is 26 feet (8 m) long.

MORE MIRRORS

Unlike most reflecting telescopes, the JWST has three mirrors. The primary mirror is concave, or curved inward. The secondary mirror is convex, or curved outward. If the telescope had only these two mirrors, the images would look blurry and stretched out. The third, or tertiary, mirror works like a person's eyeglasses to correct the image. When light is reflected off the tertiary mirror and onto the JWST's instruments, the image looks the way someone would see it if they could look directly at it in person.

This picture shows engineers in the process of building the JWST.

A CHALLENGING TASK

Building the JWST was difficult even for NASA's brilliant scientists. Before the telescope was launched, the scientists had to figure out how to solve problems that were likely to come up in space. One of these problems was how to keep the JWST's mirrors cold. Infrared light is the energy that heat gives off. If the JWST's mirrors were also warm, the heat from stars and planets would mix with the heat of the mirrors instead of being reflected.

To keep the mirrors cold, NASA sent the JWST far away from Earth so the telescope would not pick up our planet's heat. Additionally, the JWST has large shields that protect it from the heat of the sun. Together, these actions keep the JWST's mirrors around -388 degrees Fahrenheit (-233 degrees Celsius).

CHANGING TIMELINES

The JWST was first scheduled to launch in 2007. Solving all the problems that came up while the telescope was being built pushed the launch date back. When it still wasn't finished in 2011, Congress came close to pulling the project's funding. Luckily, this never happened, and the JWST's launch date was moved to March 2021. The COVID-19 pandemic caused more unexpected delays, pushing it back to December. Bad weather delayed the launch a few more times, but it finally happened on December 25, 2021.

This picture shows how the JWST flies through space with its heat shields always pointed toward the sun.

FIXING PROBLEMS

Even though NASA scientists did their best to fix problems before launch, they knew things wouldn't be perfect once the JWST was in space. Since its launch, the JWST has had several issues. For example, in December 2022, a computer **glitch** caused the telescope to repeatedly go into safe mode. The computer was mistakenly reporting that the telescope was pointing in the wrong direction, so it shut down all its instruments to prevent damage.

Over the course of two weeks, the JWST went into and came out of safe mode multiple times. Because the telescope's computer can be controlled from Earth, scientists were able to find and fix the problem. Other problems have either also been fixed or have not been serious enough to permanently break the JWST.

NASA KNOWLEDGE

In some cases, NASA was able to **predict** future problems that would not be fixable from Earth. Scientists did their best to make plans for those. For example, the JWST's heat shields are stitched in such a way that if something ripped a hole in part of one, the shield would not tear further and would still work.

NASA scientists are always checking to make sure the JWST is working right.

JWST'S DISCOVERIES

In its short time in space, the JWST has seen many amazing things. By studying the pictures and measurements the telescope sends back to Earth, scientists have been able to learn a lot of new information about space. For example, in 2024, the JWST found out more about a planet that might be able to support life. This planet, which NASA named LHS 1140 b, is almost twice the size of Earth. The JWST showed that it may have water and an atmosphere.

The JWST has also met the goal of helping scientists learn more about the past. For instance, in 2023, the telescope took pictures of two galaxies as they looked when the universe was less than 1 billion years old.

This picture taken by the JWST shows several stars and many different galaxies in deep space.

Sometimes the information scientists learn gives them more questions than answers. In 2019, Hubble's measurements showed that different parts of the universe are expanding at very different speeds. Scientists thought this might be a mistake until the JWST returned the same information. This changes everything astronomers thought they knew about the expansion of the universe.

The JWST has also created other mysteries for scientists to investigate. For example, the telescope took pictures of galaxies as big as the Milky Way that formed only a few million years after the Big Bang. Scientists say that these galaxies are too big, and some of the stars inside them are too old, to fit with what they thought they knew about the age of the universe.

NASA KNOWLEDGE

For years, scientists have believed the whole universe is expanding at a rate of 46,200 miles per hour per million light-years, or about 67 kilometers per second per **megaparsec** (km/s/Mpc). Now they know some parts are expanding much faster.

NEAR-INFRARED LIGHT

MID-INFRARED LIGHT

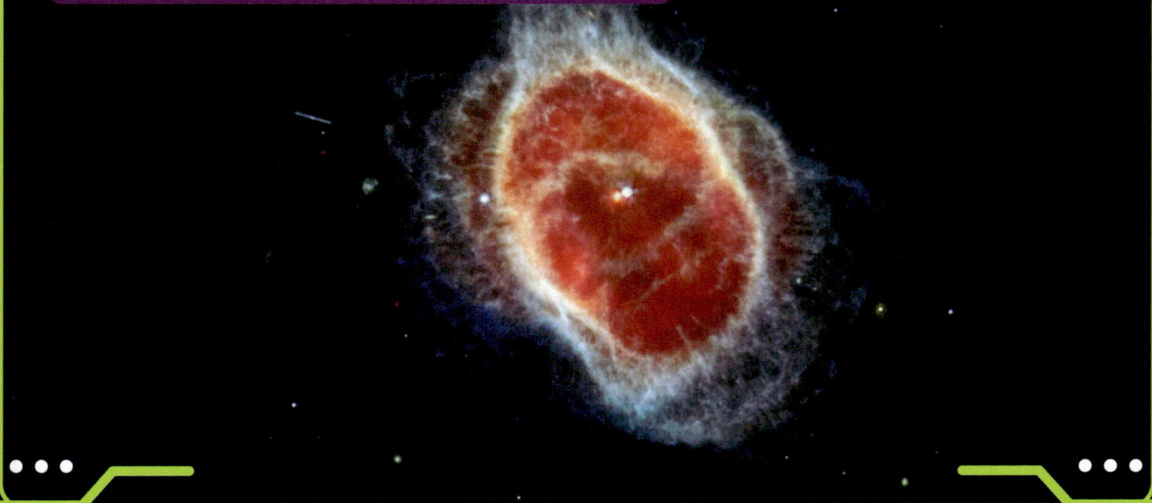

The JWST has several different kinds of cameras on it. This picture shows the Southern Ring Nebula two different ways: in near-infrared light, which is close to visible light, and mid-infrared light.

THE EXOMOON HUNT

In 2024, one of the JWST's focuses will be on looking for exomoons. These are moons that exist outside of our solar system. Exomoons are hard to find from Earth because they are very small and dim. They can only be seen when the planet they orbit crosses in front of the closest star, blocking its light.

Finding an exomoon is tricky because it means astronomers need to be looking in exactly the right place at the right time. It's so tricky, in fact, that no astronomers have proven without a doubt that they've found any exomoons so far. Astronomers hope to use the JWST's instruments to capture an exomoon's infrared light. They're focusing on a planet called Kepler-167e, which they believe is likely to have one or more moons orbiting it.

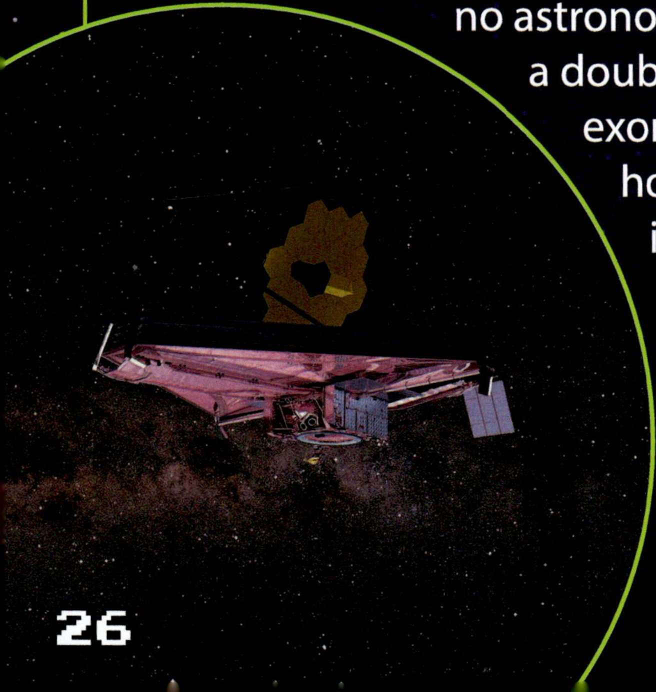

GETTING A TURN

Different astronomers have different things they want to study. Many of them want to use the JWST to help them with their research. To decide who gets a turn with the telescope, teams of astronomers submit a proposal, or an outline of what they want to study and how they will use the telescope for that research, to an organization called the Space Telescope Science Institute. The organization picks the teams with the best proposals. In 2024, 253 teams were picked to use the JWST over the next two years.

New moons are being discovered all the time within our solar system. In 2020 alone, astronomers found 10 moons around Saturn that they had not known about before.

LOOKING FORWARD

Astronomers will keep using the JWST for as long as it works. However, they are already thinking about what comes next. Hubble orbits Earth, close enough for astronauts to do maintenance work on it. This is part of why Hubble has lasted so long. However, the JWST is too far away for that. NASA expects it to last at least 10 years without regular maintenance. Although they hope it will last longer, scientists from NASA and the European Space Agency (ESA) are already working on a new space telescope.

The new telescope is called the Nancy Grace Roman Space Telescope. While Hubble and the JWST focus on small parts of space, Roman will have a much wider field of view. Roman is expected to help astronomers fix their understanding of the universe's expansion as well as find even more exoplanets.

NASA KNOWLEDGE

Nancy Grace Roman was NASA's first chief of astronomy and the first woman to hold a leadership position at NASA. Her work helped make the construction of the Hubble Space Telescope possible.

Like the JWST, Roman will be an infrared telescope. Shown here is part of the new telescope.

GLOSSARY

distort: To twist out of a natural, normal, or original shape or condition.

glitch: An unexpected, usually minor problem, especially involving computers.

identical: Exactly alike or equal.

megaparsec: A unit of measure for distances in intergalactic space.

nebula: A cloud of gas and dust in deep space.

predict: To guess what will happen in the future based on facts or knowledge.

spectrum: A continuous range or series.

technology: The method of using science to solve problems. Also, the tools used to solve those problems.

tripod: A three-legged stand.

wavelength: The distance from trough to trough or crest to crest in the line of advance of a wave.

FOR MORE INFORMATION

BOOKS

Bolte, Mari. *Mysteries Revealed.* Ann Arbor, MI: Cherry Lake Press, 2024.

Reeves, Diane Lindsey. *James Webb Space Telescope: A Peek into the First Galaxies.* Minneapolis, MN: Lerner Publications, 2024.

Slade, Suzanne. *Unlocking the Universe: The Cosmic Discoveries of the Webb Space Telescope.* Watertown, MA: Charlesbridge, 2024.

WEBSITES

BrainPOP: Telescopes
www.brainpop.com/topic/telescopes
Watch a short movie about the history of telescopes.

Kiddle: James Webb Space Telescope Facts for Kids
kids.kiddle.co/James_Webb_Space_Telescope
Learn more cool facts and take a look at some amazing photos.

NASA Spaceplace: James Webb Space Telescope
spaceplace.nasa.gov/james-webb-space-telescope/en
NASA explains more about the JWST.

Publisher's note to educators and parents: Our editors have carefully reviewed these websites to ensure that they are suitable for students. Many websites change frequently, however, and we cannot guarantee that a site's future contents will continue to meet our high standards of quality and educational value. Be advised that students should be closely supervised whenever they access the internet.

INDEX